T.I.M.E

Things I Must Experience

How to Manage Your Time More Effectively So You Can Do More of What You Love

Table of Contents

Introduction

Do you remember the story about the ants and the grasshopper? It describes a grasshopper that sang all the way through the summer, while the ants were gathering food for the winter. When the winter cold reached the forest, the grasshopper had nothing to eat and had to beg the ants for food. The ants told it to sing the winter away.

It's a very old story with several interpretations, but we tell it to our kids to teach them how planning and time management affect our future and our productivity. It's a simple concept that describes how the actions we take in the present influence the outcome of our goals.

As we grow older, we like to indulge and treat ourselves to the little pleasures of life, like eating candy for breakfast or before lunch, using all our free time to play video games or stay up late and watch TV. As time passes by, these little pleasures can take over our life, without us noticing or recognizing them as real threats. They become our habits and start to dictate our life. When a deadline pops up, we don't think of the time we wasted ineffectively. We only wish the day lasted longer.

Why you need to be organized to be creative

Many people hide behind the theory of creative chaos and how it leads to great achievements. But if you care enough to research the theory, you will find out that it doesn't apply to your everyday life. It applies to a universal structure of a pattern much greater than what your mind can perceive and imagine, so what seems like a disorganized and chaotic pattern now, is actually a great plan that will unfold sometime in the future.

In your every day, human existence chaos is just what it is – a state of utter disorder and confusion, leading only to more chaos. Let's put the theory in practice. Your desk is flooded with papers, documents, pens, erasers, and maybe even the book you read is peaking underneath some week old newspaper. If a coworker calls and asks for a document, you start looking for it on your desk, inside your drawers, and end up printing it out, just to avoid going through the mess. You didn't take the time to create different folders on your internet browser, to keep different types of bookmarks organized, so you try to think of the sites' names and search it through your history. Your email is another story. Luckily, now you can just type a keyword in the search box and all emails with the keyword will appear. But there were times when you could look for a certain message for over an hour. Searching for a file on the computer is a quest for the most courageous and those with free time on their hands.

All this chaos only creates stress and headaches. At the time, you may convince yourself that you are in control of your time and space, and that you have everything you need handy and where you can see it. If you summarize, you will realize that you spend an hour every day looking for stuff. That's at least 5 hours each week. Don't forget that these habits rub off on your personal life. You do the same in your kitchen, closet, hallway and we all know you have a chair designed to hold all the clothes you changed throughout the day, or week. Paying bills, scheduling appointments and organizing activities for your free time is also done at the last possible moment. No wonder you have no time.

As human beings, we tend to excel in finding and creating patterns. If our nature was designed differently, the ancient man would've extinct long ago. We don't live in caves anymore and we don't hunt wild animals for food, but our contemporary lifestyles awarded us with modern problems. Nowadays, if you keep your environment cluttered and you let time pass by without taking the effort to manage it effectively, you will upset the equilibrium between your personal and professional life. This equilibrium must remain balanced. Otherwise you are risking stress related diseases, professional failures and a number of limitations in your personal life. So if you have a pattern of letting the chips fall where they may, maybe it's time to change it and learn a new one. One where you are in charge and in control of your time, and your life.

Planning is Key: Keep Lists and Use the Tools at your Disposal

A wise man once said that the plan is nothing, but planning is everything. We all have big plans about something. It's a great way to gain some perspective of your life's objectives. But what many of us don't realize is that you need to determine the path of the plan, otherwise it may become a pretty picture to look at, something forgotten in the back of your mind.

Keeping lists is the best way to stay on the right track. The path to a certain goal may seem clear at the time, but suddenly you see yourself in the middle of the plan, jam-packed with paperwork, additional work and people asking you question, that require time and further effort on your part. All these obstacles can give you headaches and distract you from your goal.

Start making lists. Not just to-do lists – all types of lists. Your diary is a type of list keeping habit, so is the grocery list on your refrigerator. You know how handy your diary is when you are trying to remember when the last fight with your friend was, so let's show you different types of lists, so you can decide which will become your lifesaver:

Shopping Lists – these are necessary in every household. You will want to write down everything you need to buy the moment you take the last out of the refrigerator. Have a board in your kitchen, to pin stickers on, or keep them in a pile on the fridge itself. In time, you will know which products you need to buy weekly and which are your monthly purchases. You can make tables on the computer, containing the products you buy weekly, and just add the new products under the weekly ones.

Another stressful and time-consuming hassle we often encounter is gift shopping. Christmas, mother's day, father's day, you name it. Who's to say you can't plan gifts in advance. Keep your notebook handy in case you come across something you think someone if your life might like. You may not remember all the gift ideas you've bumped into throughout the year, so when a big event like New Year's or a birthday comes along, you can just take your notebook out and use is as a cheat sheet.

Check Lists – these types of lists are convenient when you have to arrive prepared to a meeting or event. Stress can often make you forget, so keep a record of the essential things you need for a certain interaction or occasion. Let's say you have a presentation. Your list should include all the things you need to make it successful. These things may include:

- Copies of the project's outline
- Model of the product
- Laptop, slides, projector and etc.
- Spare battery or a power cable
- Pens: red, blue, black
- Lucky charm

Resource Lists – if you find a source of information to be statistically accurate and in accordance to your requirements, keep it categorized. We live in the new century, so take advantage of the tools at our disposal. You can create several folders in your browser, all titled differently (e.g. recipes, news, fun, movies and etc.) so you can speed up the search when you most need it.

Many browsers have the option to create several users. You can name one "Work", for when you're working; another one titled "Home", for when you're surfing for fun; and of course, you can open a different user for each person in your household that uses the same computer, so your browsing history and analytics don't intersect.

It will be of great benefit if you open a Google email account just for work. That way, when your boss needs a log of your working activities or asks a question about the source of the information, you can just log in to your account and search for the sources you used. If you are the boss, ask all your employees to use the email address you provided, so you can keep track of their performance.

List Of Ideas – we come up with ideas throughout the whole day, but usually remember one or two, or none, in some cases. There are many times when you encounter a situation that inspires you to try a different course of action. Sometimes it may be an idea to check out some detail from a different perspective, or the word you were looking for just popped up in your head when you were washing your hands in the toilet. Write this little stuff in a notebook and look over the list daily.

Money Lists – this is a tough one, especially if you lack organizational skills. Keeping a list, where all your expenses will be recorded, will give you a sound evidence of where your money goes. To do this successfully, keep the receipts of everything you buy. If you forgot your receipt, write the item down on your phone or a notebook.

At the end of the month, when you calculate the totals, you will see how much you spend on food, bills, leisure and taxes. In time, you will learn what is necessary and what can be left out. These types of lists shift the focus on the big picture and will give you an idea where and how you can cut back.

Lists for goals – these are the most complex lists and call for special attention. You will need to go over each separately, and calculate the time and means you will need to achieve them. Having a goal and building a plan to get to it, will motivate you and activate all your capabilities. Without one, you may end up squandering your efforts and time, to build someone else's dreams and achievements.

According to the Zeigarnik Effect, human beings tend to remember the stuff they didn't do more than the stuff they already completed. That's why it will be best to keep a notebook ready in your phone or purse, for both the stuff you already did and the details you believe you might forget later. The first one will help you to keep track of your achieved goals. As the Zeigarnik effect implies, we often forget the tasks we finished, so it will be a nice reminder of everything you've completed throughout the day or week. Some days are so busy, you are not sure if your head is still on your shoulders, yet you feel like you haven't accomplished anything. Keep your finished tasks listed and you'll be surprised how much work you are actually capable of.

Many tools are available if you only make the effort to plan ahead. From the old-school pen and notebook record keeping, to advanced mobile applications to make your life a lot easier. There is even a specially designed device for making lists and keeping contacts on the go, called a personal digital assistant or PDA. Nowadays it's not very popular, because mobile phones are capable of supporting much more advanced applications.

Today's applications and virtual assistants have many options. You can assign different colors to each type of task in the calendar, so you can find them without having to read each. For example, assign red for meetings, green for events, blue for making phone calls or sending emails. The application might be capable of grouping different types of tasks by priority, category or date. Most have the option to remind you before a task and you can assign the actual time when the application will remind you, e.g. 3 hours before the deadline, a day, a week and etc. Best of all, today's devices have access to the internet, so don't forget to use Google Drive, Dropbox and similar platforms to store and share files, documents and photos whenever you need to.

It will be smart to make all types of lists and use several types of assistants. Download a virtual assistant for your to-do lists throughout the day and tasks related to the job. Have a notebook in your purse or pocket for the small things you come up with during the day. Put up a board next to your fridge and keep track of the spent items. Use the recording option on your phone to list the things you can't write down at the moment. We don't count those few seconds while we write something down, but when the time comes, you will realize that you saved valuable time, a time you can spend enjoying your life more.

Set Goals and Deadlines

Neurologists and psychologists agree that setting goals makes the brain consider the set objective as a reality. This means that once you set your mind on something, it has no other choice than to lead you to the goal. Often times, these goals remain far out of our reach because we fail to plan and get organized. We end up tense, because our mind is confused. We presented it with one reality, but we ended up not living it. The easier and more obvious the path to the goal is, the more nervous and anxious our brain gets. Because we are not following its directions and signs.

The S.M.A.R.T. Goals

There are basic guidelines you can follow to make sure you set the right goals. Achieved goals are like narcotic drugs for the brain. When they are not yet achieved, it releases stress hormones, to provide your body with a burst of energy and keep you alert and focused. When you achieve the goal, it releases dopamine and other nurturing hormones that make you feel naturally high. Follow the SMART criteria:

Specific – are your goals specific enough to be easily achieved? Ask yourself three questions: what, why and how. What is your goal, why you need to achieve it and how can you achieve it. If your goal is to learn to manage your time, you need to learn to manage it, so you can have a more rewarding personal life and you will achieve it when you work out a plan to save a precious hour every day. Your goals have to be specific; otherwise you will not be able to make a rough idea or an outline to get to it.

Measurable – are your goals measurable? Ask yourself how many and how much. How will you know if the goal is achieved? If your goal is to save some time throughout the day, ask yourself how many free hours you need in a day. How many minutes are you wasting while standing in line, going up and down to different floors or talking at the water cooler. How many minutes can you save, if you make your own coffee instead waiting for a cup in coffee shops and etc.

Attainable/Assignable – is your goal attainable or assignable? Can you do it by yourself or do you need to assign it to someone to help you? Do you believe others will agree to work according to your agenda? If your idea is to save time by giving your tasks to someone else, you may be setting yourself to disappointment, because your coworkers may not agree to do your job or do a mediocre one, which makes this plan un-attainable and un-assignable.

Realistic – are your goals realistic? Do you believe you can achieve your objectives or are they just wishes? Is there a conceivable path to your goal? Can you count the steps to your goal? Do you realistically believe you are capable of taking every step on the way, regardless of how steep?

Time bound – can your goal be measured in time? How many days you will need to get to the finish line? How many hours of work will it get you to finish the project? Don't set your goals for "sometime in the future". Instead, set a perceptible deadline, so you can keep track of your undertaking.

Outlining your dreams with words, with written instruction of how to get to them, is the difference between a dream and a goal. You will have to stop thinking narrowly and brainstorm ideas. Don't decide on the definite course of action just yet. Take a couple of days, with your pen and paper ready and think how you can become more organized. Sometimes, it just so happens that we set too many goals at the same time, so we end up doing none successfully. Focus on the quality instead of the quantity. Focus on the little things, the details. If you work in customer services, for example, and your job is to write to users about the company's services, it will be time consuming to write the same emails all day long. Instead, take an hour and work on a couple of templates. That way, when a client asks for information or reports a problem, you will have a courtesy letter prepared and all you'll need to do is focus on the problem the client is having and addresses it.

Avoid multitasking unless you have mastered it. Trying to complete several tasks at once may hamper productivity. Test your skills and try to perform several chores at once. If you are headed in a direction where later you'll also need to make copies and talk to a coworker, by all means, do it. It will save you time. But in other cases, where important tasks are at stake, it will be better to focus and concentrate. You may think you can do it all, and you probably can, but you have to take into account external obstacles, like changes in directions or coworkers dropping by or asking for help.

Follow your own direction. For a plan to work, it needs to be thorough. You need to plan and calculate each step. That way you can estimate a possible completion date and feel good about the goal before the actual deadline. Let's say you want a promotion at work, and you give yourself 3 months to get it. Here's what you can write down:

Lisa's Promotion Plan

Daily tasks:

- Be nice to coworkers
- Help out
- Don't whine
- Don't complain
- Don't gossip
- Be prepared
- Come in half an hour early
- Stay half an hour late
- Get noticed by the boss and coworkers

Weekly tasks:

- Finish tasks before deadline
- Present your results
- Do more than you are asked
- Mention related accomplishments to boss – sell yourself
- be dynamic at meeting

Monthly tasks:

- Learn a new program, or skill
- Ask for more duties
- Be a team player
- Brainstorm different approach, ideas, tips

At last, don't talk about your goals and plans to anyone. You can never predict other people's response and reactions, which may hinder your feelings and motivation. Stay quiet and positive. Be flexible about your plans. Remain alert and with both hands on your plans. If something goes wrong unexpectedly, review the steps that led to the failure. Maybe there was a flaw in your plan you overlooked or miscalculated. There's always something you can do to repair the route to the goal, as they say: if there's will there's a way.

Prioritize Work That Is 'Important but Not Urgent

So you've written down your plans. Now what? Most people will choose the easiest task to complete first, to feel good and accomplished. Picking the easiest way out is called the path of least resistance. Mark Twain said if you eat a frog in the morning, that's the hardest thing you'll do all day. Once you complete the hardest, most complex of your tasks for the day, the rest will be easy.

Which task is most difficult or unpleasant? Is there a task that will take most of your time? Can you split it in several smaller tasks, or do you need to do it all at once? Imagine you have only one closet, for both your summer and winter clothes. Every spring you have to pack the winter clothes in bags and take out the summer clothes.

If you take out the winter clothes and the summer clothes on the bed to refold or wash them, you'll have a huge pile of clothes in your bedroom. You'll never know where to start and you'll probably waste a lot of time sorting things out. But if you take the winter clothes out of the closet first, fold them and place them in bags, and only then fold and sort the summer clothes to place them in the closet, you've done it right.

The tasks that make the project are the same: sorting, folding, packing. But the order in which you complete them makes all the difference. In the first case you take the path of least resistance: taking all clothes out on the bed to avoid wasting time sorting them and folding them at the moment. When everything is a mess, you sort and fold anyway. Only now you waste time on other details, like cleaning after yourself and making sure the right shirt is in the right pile.

You have to be methodical and think strategically. When you decide you need time management in your life, you will probably want to change your pattern in several areas at once. You can do it even though it may seem overwhelming, and it will be for the best too. If you focus on your career too much, you may undermine the bonds in your private relationships. Prioritize your plans. Use Stephen Covey's system:

- **Important and Urgent**
- **Important and Not Urgent**
- **Not Important but Urgent**
- **Not Important and Not Urgent**

This system will help you decide the plan or path to your goals, the steps and their order. In Lisa's example, important and urgent goals were to come in early and stay in later, get noticed by the boss and coworkers and finish tasks before the deadline. Important but not urgent were: be dynamic at meetings, present results, learning a program and etc. Sit down and think of a strategy that will work best. If there are too many chores in one category, examine them exclusively and focus on the timetable for each.

Another way you can weigh the importance of your goals is by deciding which will be pursued monthly, early, by the end of 2 years, 5 years, 10 years and etc. When you give yourself a deadline, it's important to not wait for the last possible moment to act on it. Split your big plans into smaller and smaller parts. Let's take Lisa's example again. Even though the "learning a program" goal was in the monthly list, she can still list "buying it and installing it" in her weekly list. She can try to find tutorials in the meantime and join forums to research the options, pros and cons of the program. She can learn about a better program and offer a solution when her boss complains about their current program, and complete 2 goals of the plan: get noticed and brainstorm ideas. Sometimes it's better to make several smaller steps instead of one big one, because regardless of how small they are, they will still take you to your goal.

Utilize a different system to prioritize your tasks. List them under: tedious, less tedious, bearable and done in a minute. Eat the frog early in the morning and the rest of the day will be easy to manage. Your old train of thought might lead you to believe that the highly important tasks should be done last, after you complete all other small activities. But most of the time, these small activities can wear you down and leave you no energy for the other chores. When you are done with the big stuff, regardless of how exhausted you are, the fact that you have only small errands left, will be the boost you needed.

Ring-Fence Your Most Creative Time

Scheduling the right time for each task will define the outcome of your plans. You know yourself best and you are familiar with the ups and downs you experience throughout the day, whether in mood or energy. There are two types of people: morning people and evening people. Most morning people wake up without touching the snooze button. They are peppy up until lunch time, when you can notice the first signs of exhaustion. Evening people are merely useless in the morning, especially the first few hours of the day, but once they get the ball rolling, no one can stop them. People from both groups are capable of finishing the same tasks in the same given time.

According to Rose Zacks and Mareike Wieth, and their study, creative work and brainstorming are best done when people are not at their peak energy levels. So if you are a morning person, it will be best to complete the chores that require more energy and concentration early, then focus on the intellectual, artistic and emotional aspects of the job later in the day. For evening people, it's the opposite. They will feel more productive, quietly planning strategies and brainstorming ideas in the morning, and work on the demanding tasks later.

It's not an exact science, at least not for everyone, so you'll have to monitor your activities to understand better when the right time for each activity is. You will find out that your blood pressure drops after a big meal and you feel a burst of energy an hour after a workout. Your diet may also alter your energy levels, so it will be best to snack on carbohydrates around lunch time, so your body can receive new energy for the rest of the day. Avoid carbohydrates before bed, because your body won't be able to use them as energy and will store them as fat.

Consider distractions and interruptions in the equation. When is your working space most cluttered and busy? When are coworkers likely to come in unannounced? When your brain is in an alert state, in its peak period, distractions and interruptions are easily processed and filtered. When you are tired however, your concentration and focus can be easily influenced. Chose the time of the day when you can focus without much effort. We have three periods in the day of around 90 minutes each, when our brain is capable of following a train of thought without losing concentration. That's more than 4 hours every day for the most demanding chores.

Not many people understand this body rhythm of ours, and want to do all challenging tasks at once. The brain has its own rhythm and it works in cycles. Beta and gamma waves are the impulses that steer our brains to think, learn, memorize and focus. However, the brain can't keep these waves dominant for a very long time, or it will become overstimulated, exhausted, and stressed. So instead of working 4 hours straight on the demanding parts of the job, thinking you will take full advantage of your brain capacity, follow its natural cycles. Your brain will send signals when it's time to relax, by giving you a headache or recurrent distracting thoughts.

It will be pointless to go against this natural rhythm, because you can't change it. You can influence it with stimulants like coffee or drugs, but when they wear off, the slow and relaxing waves will take over again, making you even more tired. Use this natural rhythm to your advantage. Theta waves, for example, are valuable impulses that stir creativity, intuition and emotional connections. When the alpha waves hit the brain, you are most capable of problem solving, thinking outside the box, summarizing and visualization.

Make a plan to go over an important task during a gamma or beta brainwave state, a chore that requires concentration, numbers and handling large amounts of information at once. You have less than two hours until your brain starts sending signals that it can no longer concentrate, which means the theta waves took over the cognitive processes. Now it will be a good time to use your right brain hemisphere for creative work, intuitive matters in question, or tap into the spiritual states of the mind. Alpha waves are mostly used for distressing and relaxation, and scientists discovered that just by closing your eyes for a few minutes, your brain will start generating them.

Avoid the 'Sisyphus Effect' Of Endless To-Do Lists

So you've decided what you need to work on and the lists got a little overwhelming. So overwhelming in fact, that you lost all the motivation to get started. You wonder how are you going to do this and where are you going to find the time. This happens very often and it happens to all of us. When you first start making plans, you calculate the time and effort you need for each task. At the moment they don't seem farfetched and overwhelming, because you examine them individually, but put together in one gigantically long list they seem impossible. Relax and take another look at your notes.

When you have 15 things written down for one day, it's pretty easy to get terrified by the amount of responsibilities and duties. You don't need to be because there are probably 5 things you can scratch off as unimportant. Scratch those things off, keep the original and rewrite your to-do list. Now you'll be left with 10 tasks. 10 tasks in 8 working hours don't seem overpowering, but then you remember that coworkers will be loud, or ask you for help or direction, or chat by your desk. Maybe your boss or client will want to talk in the middle of your power-hours and you realize that your to-do list is still a little cluttered.

It seems like there's no way out of your busy schedule, but there is. Make a new note for tomorrow tasks and write down only 5 tasks. Again, think of the type of person you are and the type of activities you are capable of performing fast and accurately, with distractions and under pressure. Do those things when you are at your peak, when the day starts, when everyone is looking over their objectives for the day, asking for assistance or agreeing on a directive. When you are done with these 5 tasks, you can look over the 10 task to-do list. Choose the ones that can be completed immediately or that are moderately urgent. Pick those tasks that can be performed in a single effort, or in the same location. Suddenly you'll realize that you are thinking of completing tasks from your spare lists. It should give you the satisfaction that you are ahead of your schedule, instead of behind.

Get Things Done by Putting Them off Till Tomorrow

Everyone always says tomorrow is not promised to us, and that we shouldn't do tomorrow what we can do today. But when your schedule is packed full with activities, tomorrow seems like a perfect day. At least for some activities. That's actually the secret to a better time management. You have to make a list of all the tasks you are capable of finishing, prioritize them, and perform them according to their urgency. When the time comes for the easiest duties or those that can be postponed – put them in tomorrow's to-do list.

That's the difference between promising to do something tomorrow and postponing it until tomorrow. When you are putting stuff off for tomorrow, you are making a plan to do them. Something will be missing if you don't do them and you won't feel accomplished. Write down all the chores you have to postpone, keep that list separate and jump on those errands the first chance you get. That should be your priority. If you postpone 4 out of 5 tasks every day, it's no longer putting off, it's procrastination.

These tasks should be something minor, something that wouldn't crook your path. Big, important tasks cannot be put off, even if they are overwhelming. In that case, you can split your works in several days, but never postpone it. Experts make a difference between active procrastination and laziness. When you are actively procrastinating, you are not just postponing stuff for later. You are going through your other tasks and looking for the next one to cross of the list. You are still actively working on a plan, only delaying the unimportant and non-urgent tasks

Postpone the small stuff. If there's a task in your tomorrow's list that can wait until the next day, assign it with a question mark. When coworkers ask you for assistance, ask if the call is urgent. Learn to say no. An email pops up and you see it's not urgent, reschedule the reply. Unless you work alone and have no one to keep updated, chances are great that your working hours are filled with distractions, unexpected and additional chores. Learn to identify their urgency the minute they are assigned.

It's all about the flexibility of your job and your character. Study your environment, your coworkers, and the dynamics at play. Avoid the rush-hours in the working place. If Friday is a busy day for the copier, copy your papers on Thursday. If accountants work full steam ahead on Mondays, pick another day to ask for advice. Discover a system where you can avoid most distractions and make things come about without pushing them hard.

There's the right time for everything and being able to put things off for later is a proof of that. You are never given more than you can handle, and if you think you are, it's only to get you to see the situation from another angle. To teach you to be more organized and work under pressure. If someone's telling you their job is sunshine and rainbows, they are probably not doing it right.

Get Things off Your Mind

Now, this is a tricky one. Getting things off your mind is a little easier said than done. You know how sometimes you get that one song stuck in your mind all day. It starts with a few lines in the back of your mind. Then you start singing along and realize you are singing along, so you stop. It gets a little boring and annoying hearing the same song for hours, but you also can't get rid of the echoing melody. Our brains are powerful machines that can carry out complex processes, yet they tend to get stuck on the same lyrics or tunes all day. Next time you can't stop singing a song, try finishing it. That's how your brain works – it needs to finish the given task or it will remind you about it until you finish it. It becomes a broken record that will keep singing the same tunes until you lend it a helping hand, and help it move on. Once the train of thought gets going, it has to reach its final destination. So, give it a chance.

The working place is usually full of distractions, and focusing requires tremendous effort. When you need to finish a certain task without distracting thoughts and people messing up your process, you have to build your environment in a way that it doesn't allow interferences. Put up your schedule where everyone can see it. Put your visiting hours in there. A time of the day when coworkers can ask for stuff and a time when they absolutely can't. If you don't have the freedom to schedule such hours, use other means. Download alpha brain waves (binaural) beat, and put on your headphones. Log off of any social platforms, minimize the unnecessary windows and set your phone to silent.

Deal with distracting thoughts the same way you can deal with the repeating tunes. Let them in, examine them, allow them to take their course and let them out. If you keep blocking them without trying to resolve their occurrence, they are just going to come back in. Sometimes, thoughts are not easily controlled, and it seems like our minds have a mind of their own. In that case, ask yourself a couple of questions that can help the distracting thoughts move along faster: Is this more important? Is it the end of the world if I don't think about this right now? These are my working hours, what am I doing, thinking about my cat?!" Rationalize your thoughts and they will lose their meaning. Close your eyes for a minute and focus again.

Review Your Commitments

We always neglect our New Year's resolutions because we forget about them soon after the holidays. The same can happen when you don't pay attention to your plans and tasks that make them. Review your commitments every morning, every Saturday, every 1st of the month. Review all your plans and goals: your dream boards, your working schedule, your free time plans, every list you ever made. Keep your done and finished lists in a pile somewhere, or take their photo, to remind you how well you did and how hard you worked to accomplish them.

Reviewing will help you see the big picture. Probably every morning you start to feel a little anxious about your day. Sit down with your list of tasks in front of you and start visualizing how you perform your duties. Researchers have done experiments to prove that the brain shows the same activity, whether you perform a given task or you are only imagining it. Imagine yourself doing all the work. If there are too many things on your list, imagine yourself doing the first three successfully. You can consciously raise your energy and enthusiasm to perform all the tasks, when you can already see yourself doing them in your mind.

Reviewing is a necessary action for proper judgment. At the time of the making of the lists, you may become overeager and convince yourself you can accomplish the inconceivable. Reviewing will keep you grounded. In time, when you get trapped in the middle of the road with no apparent way out, reviewing might open your eyes for more solutions. You may realize you've been doing something wrong, or that you can do something faster in a different way than the planned. In any case, it's good knowing you have something to rely on.

Resources to help you get things done

Our modern lives are complicated, it's a fact. Among the many choices we are offered every day, we have troubles picking the right things. We are often swamped with information we don't know what to do with, so we end up letting it slide through our brains without properly processing it. Fortunately, there are people who think about our productivity for us.

Evernote, Dropbox, Quip, and Onenote are all apps designed for taking notes, storing favorites and sharing. You can use them to save documents, bookmark searched information and collect music. You can easily find your files by using keywords in the searches, and of course, by tags. They are free for limited use, but if you need a lot of space or more updates per month, you can become a premium member and enjoy the benefits. You can download them on all your devices, and have everything important easily accessible.

Todoist and Asana are task management application you can use besides the calendar on your phone or Google. Todoist design is clean and simple, but it allows many features, like collaboration between users, reminders based on location, great classification tools and offline functionality. Asana is the rising star app among managers, because it allows many features like establishing a plan through creating tasks, commenting on the created tasks, mentioning other users in the comments, assigning them to a task or as task followers. You can add deadlines, search related tasks and receive notifications when other users completed their task. It's a user friendly platform that allows full communication with all parties involved in the project, only through the app.

If your job involves the social media platforms, you can choose between Hootsuite and Bufferapp. With Bufferapp you will no longer have to worry if the scheduled information reaches the World Wide Web in time, because you can appoint the time of the day when you want the app to post in your name on several platforms including, Facebook, Twitter, Pinterest, Instagram, and LinkedIn. The con of using Hootsuite is that the user is not allowed to control when the app posts the info, but they send email reports on how each post is doing on the internet. The pros are that it allows you to use more profiles and unlimited number of posts, where Bufferapp allows only 2 profiles and 10 posts per profile every day, assuming you have a free account.

These are only a few of the most popular applications many business oriented people use, to stay productive and organized. You can research more tools and see which one is the best for you and your professional or personal planning strategies.

Conclusion

In the world that we live in, organizational skills should be taught at school. Communication skills, self-defense and manners should be taught also, but that's a different story. The point is: without these skills developed, you will not be able to function as an adult. You may struggle to survive and make ends meet, whether in your social or professional life, but the point of life can't be to just survive.

As modern people we carry a lot of stress on our shoulders. We need to think about work, our kids, our social lives, what we say, what we do, what we post. If you are not organized, your work is going to interfere in your social life, your social life will go idle, everyone will be disappointed because you are never 100% there, and you'll live life under the pressure of uncertainty.

All that can be changed if you start making plans. Big plans, small plans, instant plans, long term plans, make a whole scheme of plans! At one point, we all get caught up in the moment and lose our ground, but once you have a note to push you in the right direction, you will turn out all-right.

www.ingramcontent.com/pod-product-compliance
Lightning Source LLC
Chambersburg PA
CBHW071555170526
45166CB00004B/1687